ΛLIENS ΛND UFOs

by
Rick Bennette

ISBN-13: 978-1500398101
ISBN-10: 1500398101

Published by BeelinePublishing.com
Tequesta, Florida

Dedicated to my loving wife, Denise

Preface

When you think of aliens and UFOs, do you think of little green men, or perhaps gray beings with large heads and big eyes? It makes for a nice science fiction movie, but the reality of what aliens are, if they are, is likely to be much different than you think.

In this book, we will examine known scientific facts and apply those to the most plausible explanation of alien life. The two biggest questions about aliens are these. Do they exist? And are they here with us now?

While there is a vigilant effort by scientists and astronomers to search for the existence of life anywhere else in the Universe besides here on Earth, so far only theories abound.

Here is a personal point of view driven by logic, physics and science to shed some light on these two questions. Does life exist on other worlds? Have they ever made it to Earth?

Chapter 1

Here on Earth, there are a number of people who will swear they have seen or encountered alien beings. If you have any doubt about this, all you need to do is search Google or Youtube for UFOs or aliens. You will find thousands of self proclaimed eyewitness accounts and video clips of encounters with entities presumably not of our world. Could all of these accounts be mistaken? I don't think so. But I do believe there are more plausible explanations for these sightings than most of their tellers would have you believe.

If aliens really exist on Earth, then where are they? For as many stories have been written, and as many accounts of alien sightings have been witnessed and filmed, it all seems to come down to just one thing. Do you believe?

I'm a factual kind of guy. As much as my imagination can run wild, and it can, just look at some of my other books on Amazon to prove this, I tend to follow scientific facts and especially the well established laws of physics to support my own belief system.

Based upon science and physics, my own personal belief is this. I believe without a doubt there is life elsewhere in the Universe. It's a big Universe out there. We already know there are billions of other solar systems in the Universe. Astronomers have so far confirmed the existence of thousands of other planets revolving around other stars in our own galaxy, and more are discovered every day.

Recently, several new planets have been discovered within the Goldilocks Zone of their stars, that is, the point where water can exist as a liquid. Scientists believe that water in liquid form is an essential ingredient to life as we know it, or at least life as best as we can predict. As humans, we must realize our perception of life on other worlds is likely to be clouded by our limits of exposure to the life forms known to exist within the confines of our own land and oceans.

Naturally, not every planet can support life as we know it. Liquid water and moderate temperatures are two major factors in determining whether a planet can support the kinds of life we know here on Earth. A planet must be a certain distance from its host star in order for water to exist in liquid form and have temperatures that are hospitable enough for advanced forms of life to exist as far as we know. As recently as a few months ago, new planets have been discovered with conditions favorable enough that scientists believe they could actually harbor life as we know it. Of course, the possibility exists that life forms other than those familiar to us might be able to exist in conditions too hot or cold for life forms as we know them here on Earth. All one needs to do to open their mind to this possibility is to research the accounts of life recently discovered within the intense heat of the rims of volcanoes, or deep within the cold and crushing pressures of our deepest oceans. Odd forms of life exist here in deed. But they aren't exactly the kind of creatures likely to surpass us in the race to conquer the stars.

Mars, our closest cosmic neighbor besides the moon, is the first place we're looking. But we only expect at best to find traces of microbial life forms beneath the soil of Mars using our latest rover technology. Sorry, there are no little green Martians to photograph or bring home. Conditions on Mars are already known well enough to modern scientists. The atmospheric and weather conditions on Mars prevent higher forms of life from being able to exist.

In the event we someday discover microbial life on Mars, it will open up immensely the possibility of discovering more advanced life elsewhere in our galaxy. As soon as we can confirm the existence of life in at least one other place in the Universe besides Earth, we will open up our collaborative beliefs for life to exist on many other worlds as well.

Most places in the Universe are still well beyond our technical ability to find life. The places were are able to conduct a search with reasonably accurate results is about the same ratio as comparing your quarter acre back yard to all the land masses on Earth. In other words, we can't really look too far yet, anymore than you can search within your own property and base all of life on Earth according to your finds. Even in our own tiny cosmic back yard, astronomers continue discovering new planets every day. There are trillions more they predict exist, but can not yet confirm.

The Almighty Power that created the Universe, the Earth, human beings and all the plants and animals which are so delicately balanced in nature, must have created life elsewhere as well. The precise balance of nature and all living things on Earth is too perfect for me to believe it wasn't at least designed by a Higher Power. The Universe is simply too large for me to personally believe our Earth is the only planet blessed with the presence of life. Earth is but a microscopic spec in the scheme of the Universe. Less than the smallest visible drop of water as compared with all the oceans of the world. It makes little sense that life would have only been created in this one tiny corner of the Universe.

If life as we know it also exists elsewhere in the Universe, chances are those beings could have many similarities with life on our own planet. After all, if one Almighty Creator designed and created the entire Universe, and therefore all life within it, it stands to reason there would be similarities between life on Earth and life found on other worlds. To prove the theory of Universal Creation, all we need to do is examine some facts already in evidence.

We already know there are seven other planets in our own solar system and there are trillions more outside our solar system. We have positively observed thousands of those planets, and applying what we know about them, we can accurately predict there are trillions more. We also know there are billions of stars in our galaxy. We can see them every night.

Scientists who study the stars and planets have proven these worlds are composed of the same elements as our own sun and Earth. Of course, the quantities and various combinations of those elements are not found in the same proportions as they are here on Earth. But the raw elements are indeed the same. Why are the building blocks of planets and stars light years away from us the same as those here on Earth? The most plausible answer is because everything in the Universe is created from one original source. Whether or not you choose to believe that source is God, a cosmic field of power or something else entirely, you must concede that some common entity created all that there is in our Universe, if only because it is here.

Believing in the existence of life beyond our own world is one thing. Believing that an alien life form has made the journey to Earth is something else entirely. The belief of life elsewhere in the Universe, and the possibility of alien life having made it to Earth are supported by two entirely different sets of facts.

Although I believe in life elsewhere in the Universe, the immense distances between star systems leads to difficulty in my belief that other beings may have already made it to Earth. Not to say that it isn't possible, but just to say that the possibility is so slim as to make the probability unlikely. I will, of course, explain this at greater depth in later chapters.

It is evident in my mind that many people will disagree with me, claiming that discoveries of places on Earth such as the great pyramids could not have been built by humans of that time period. It's a tough call to make without absolute, irrefutable proof, so even though I can't foresee the *probability*, I am still open to the *possibility*. To put these terms into perspective, allow me to illustrate in a manner easier to visualize. Buy one lottery ticket. By doing so, there is the *possibility* you might be the sole grand prize winner. But the one in a hundred million mathematical odds of you winning is only *probable*.

As far as the belief by some that past civilizations could not have been talented enough to create the grand structures of their day, all one needs to do is look around at our own societies.

Fifty years ago our Country was revered by most societies as the most powerful and influential nation in the world. Our technology created automobiles, powered flight, electric lights, put men on the moon and gave us creature comforts and instant communication unheard of only a few generations ago. Yet today our once mighty Nation stands in embarrassment as perceived by other countries because of our massive debt and many other problems we face today.

The Arabs at one time were looked upon as the cradle of civilization. Today, they appear hell bent on destroying most of it. So is it not out of the realm of possibility that civilizations many thousands of years ago were once great enough to create the masterpieces in their day, only to have later wallowed in their social failures and other problems?

Chapter 2

As a child, my first recollection of seeing the stars and the moon in the sky was something that changed my perception about life. I was probably four years old when I remember thinking to myself, what are these lights in the night sky? How far away are they? Can they ever fall down out of the sky?

Obviously, these questions were answered as I grew older and began to study the subject of the stars and planets.

Stars are more than just points of light in the night sky. They are large balls of gasses much like our own sun. Some are smaller, but most of the stars we see at night are a good deal larger than our sun.

Planets are those bodies that revolve around the stars. While stars contain and give off their own energy, planets typically do not. Like our own Earth, other planets rely upon their stars to receive energy, much as our own Earth relies upon the sun for its source of energy.

From the vantage point of Earth, stars appear to remain in the same place in the night sky every night, moving only as our Earth rotates from day to night and back to day. The reality is, stars are moving through space at extremely high speeds. Because they are so far away, they appear as if they are not moving at all.

It is much like watching an airplane in flight. If you are sitting at the airport, the plane appears to be whizzing by very fast as it takes off or lands. That same plane, observed later in flight by someone looking up from below, will appear to be creeping across the sky very slowly. In reality, it's actually going much faster than it does during takeoff or landing. It is the relative point of the observer that makes the speed look faster or slower.

How far away are the stars? The nearest star to Earth is a little over four light years away. The farthest point at which we are currently able to observe galaxies is a little over 13 billion light years away. Just how far is a light year? It's hard to imagine in miles, especially for those of us who have a difficult time imagining even how far away five miles is.

It is not so important to know the math as it is to try to comprehend the immense distances between the stars. This fact will come into play when predicting the possibility of aliens coming to Earth. Ready? Here come all the numbers.

A light-year is the amount of time light can travel in one year. We need a term like light-year because using miles would present numbers so large, it would be hard to even write them down and comprehend their vastness. To make the example, try imagining this. There are nearly six trillion miles in a light year. Here is that number printed out. 6,000,000,000,000. See what I mean? And that's just one light year. The Universe is predicted to be about fifteen billion light years across. Try imagining the number of fifteen billion times six trillion.

Let's break it down. In one single second, light travels 186 thousand miles. That is about 670 million miles an hour, which is about a million times faster than a jet airliner.

To put that into perspective, our Earth is 25 thousand miles around if you measure the surface all the way around the equator. This means in one single second, light can travel around our world over seven times.

At this unbelievable high rate of speed, light from the moon, about a quarter of a million miles away from Earth, takes about one and a half seconds to reach the Earth.

Our sun is 93 million miles away. Light from the sun takes about eight minutes to reach the Earth.

Light from the nearest star takes nearly four and a half *years* to reach the Earth. So we say the nearest star, which by the way is named Alpha Centauri, is about four and a half light years away. And that's only the nearest star. With our current rocket technology, it would take eleven thousand years to reach Alpha Centauri. You'd need a lot of food, water and air on board to make that trip. And a really good doctor. Obviously, we aren't headed that way anytime soon.

The majority of the stars we can see at night without a telescope or binoculars are between several hundred and several thousand light years away. This makes travel times between star systems improbable for us or for any other life forms out there, even if they have spacecraft that can travel at light speed. Wormholes, you might suggest? That's just the stuff of science fiction. You see, if two objects are a certain distance apart, there is no shortcut to making that distance less than it is without affecting the relationships between every other object in the Universe. The time-space continuum might change relative to the people in a light speed spacecraft, but the distances between stars don't change, except by their own orbital paths through their galaxies. And those speeds are nowhere near the speed of light. This is why the stars in the sky all seem to appear the same distance relative to our place on Earth. The speed of stars traveling through space is quite insignificant relative to light speed.

Why have I gone on and on about these distances? To reinforce the point already made in the last chapter. It would take a massively long time for our current technology of spacecraft to reach even the nearest stars.

Presently, our fastest rockets can move along at about 25,000 miles an hour. Using the gravitational fields of the planets in our own solar system by slinging our spacecraft past these planets can double that speed. Even at 50,000 miles per hour, it would take about half a billion hours to reach the nearest star. That's well over five and a half thousand years, which brings me closer to my point. There is no known living organism that can survive a trip that long.

If you believe in the Big Bang theory that states all of the Universe originated from one massive explosion nearly fifteen billion years ago, and consider one common Creator designed all of it, then it remains plausible that life on other worlds might be constrained to similar limitations as the thousands of different life forms found on Earth. Very few species can outlive a human, with sea turtles perhaps being one of the longest living creatures at 150 years.

Could we build a spacecraft large enough to supply food for a trip that would last nearly six thousand years? It would take a spaceship a dozen miles long to contain enough food, air and infrastructure to last for that amount of time. That's assuming you can preserve food that long, or grow more of it as you go. Obviously, the same people who left Earth wouldn't make it to the stars. There would be multiple generations of people who would live, die and repopulate for all those years. It is possible there would be many more born than originally left Earth, unless some draconian methods were agreed upon to control the re-population. Look how humans have multiplied in a nearly six thousand year span on Earth. In space, you' don't have the luxury of feeding and housing a population growth of that magnitude. We are also assuming the spacecraft itself would stay intact without life threatening failures all those years. One encounter with a rouge asteroid larger than a golf ball would obliterate the life support systems.

OK, you say, let's learn to build a faster rocket. Say,

one that can travel at light speed and get to the nearest star in four and a half years. This way, one group of astronauts could make it to our nearest star and back within one generation. Such a rocket with light speed ability would consume more energy than exists on the entire Earth today. That's just to get there. You would need to have a similar amount of energy to slow it down once you got there, otherwise you'd whiz right on by. And what's the point of that?

You would need to double that total amount of energy once again if you ever wanted to return to Earth. It isn't plausible to believe we will ever be able to build such a light speed spacecraft unless there is a monumental discovery of a yet unknown and much more potent energy source in the future. Even if we could build such a spacecraft, we still need four and a half years of food, water and atmosphere for each person, plus entertainment for all that time. Double that for a return trip.

Physicists acknowledge that the laws of physics are uniform all across the Universe. That means if humans have these insurmountable obstacles to get to the stars, any other life form out there would likely experience similar obstacles getting to Earth.

Of the hundreds of thousands of life forms right here on Earth, there is only one species capable of even thinking about such a journey. Humans. Even if there is intelligent life out there among the stars, there may not be a civilization capable of considering space travel. If there is, they may not have access to the materials or resources to build it. Humans were fortunate to discover crude oil and iron within our soil, the materials from which all things plastic and metal are created. Other planets capable of supporting intelligent life forms may not have those resources, and therefore, no amount of knowledge would ever be a substitute for the materials needed to build a spacecraft.

For the sake of argument, let's assume an advanced civilization on the nearest star Alpha Centauri could build a spaceship capable of light speed, and could stock it with the necessary provisions to reach Earth. They, too, would need a large spacecraft to contain all their supplies for a journey long enough to reach Earth. It would be a spacecraft at the very least, several hundred feet long.

Our government agency, NORAD, can track objects in Earth orbit as small as a marble. Astronomers can track meteors a few hundred feet in diameter from millions of miles in space. They can predict how close these objects will come to Earth from several months or even several years in advance. Any spacecraft that size headed to Earth could never arrive undetected.

Don't you think NORAD and astronomers around the world would track the path of any spacecraft that large long before it ever reached Earth? But, you say, the government would simply hide the fact of such an inbound spacecraft. Even if the government chose to cover up the approach of such a spacecraft, there would be some astronomer or scientist who would break the news to the media, who in this day and age, would have a field day broadcasting it. After a while, even our mighty and secretive government could not contain such an event from becoming public. The scientific facts tell us that even if there was an alien spacecraft that made it to Earth, it would not arrive here without plenty of advance warning. They wouldn't be coming in a spaceship the size of a small airplane.

Any civilization capable of reaching Earth would by necessity be an intelligent species. Should this ever come to pass, those beings would likely be more into research and discovery as opposed to hostility. That isn't to say such a civilization wouldn't be capable of violence, only that, like our own astronauts, those selected to venture beyond their own world would probably be peaceful. The signals of our radio and television broadcasts, even if they could not decipher and view our programs, would at least alert them to our presence long before their arrival. If they did have a means of decoding our signals into images they could view, they would get a glimpse of humanity long before they arrived. A glimpse, admittedly, not quite representative of how our life really is. They would likely wish to attempt some sort of communication with us, even if it was nothing more than electronic blips we didn't understand.

There will always be the doubters and the naysayers, just like those who believe the moon walks in the sixties and early seventies were all hoaxes. You simply can't change everyone's mind, even with irrefutable scientific facts.

On the other side of reality are those who already believe aliens are here. These believers are far and few between, and most people don't put much faith in the ramblings of such over zealous believers. Humans are a species who need to see evidence before we can believe. After all, there are those among us who even chose not to believe in God simply because we can't see Him and prove His physical existence. Even though we may not all agree on the name, or agree about who God really is, most of us in the world have faith in some Supreme Being.

With that faith, why is it so hard for those who believe God created life on our world to believe the same God might have created similar life elsewhere in the Universe? After all, it's a big Universe out there. Perhaps it is a Divine plan to have created the stars so far apart with the intentional purpose of preventing distant civilizations from ever coming together.

Chapter 3

Over the years, there have been thousands of sightings of UFOs. It is important to keep mind of the fact that UFO stands for unidentified flying object, not automatically an object from another world. This, of course, means that any object in the sky you can't identify is a UFO. That includes exotic looking aircraft formerly unseen by the general population designed by governments for secret missions. In fact, I'd be willing to bet that almost all UFO sightings are of this origin. Those that aren't can be explained by some other means if one is willing to conduct honest research with an open mind.

I had my own personal UFO sighting a few years ago, which I documented on high definition video. You can find this on You tube if you search Christmas UFO Tequesta. My wife and I were on our way home after Christmas dinner with the family. Rounding the turn, several illuminated objects were floating through the sky in exact formation with each other.

Based upon the flight characteristics of these objects and their appearance, the objects defied all logic of being some kind of airplane. The formation of yellowish-white lights in the night sky appeared to be moving too slowly and were rocking back and forth too steadily to be conventional aircraft. In addition, there were no red and green navigational lights, and no flashing strobes as are required for all aircraft flown in the US. Even a military aircraft would have some sort of lighting at night.

To further support my findings, after posting my video on You tube, I received emails from observers in California and Germany confirming the sightings of the same lights. I shot the video in Florida. In order to be observed from these three distant locations simultaneously, this led me to believe the lights were much further away than they appeared to be. To my eyes, the objects looked as if they were several miles away, maybe even a dozen or more. In order to be spotted as far away as our three locations, that meant the objects would have had to have been many thousands of miles away. Being that our atmosphere effectively ends at 62 miles above the Earth, that clearly puts the objects into outer space.

Another explanation could have also been the observations were made minutes apart by a high speed object in orbit at least several hundred miles above the Earth. They didn't appear to be moving fast enough for that. No airplane can fly that high, and a satellite would have had a smooth and steady trajectory through the night skies, not the rocking back and forth as these objects were clearly doing.

So what could it have been? Had I simply assumed that all three of us had seen the same thing, it would have had to have been something in space. I asked the person in California to send me their video. It looked similar, but not exactly the same as mine. It became obvious what we all observed was similar in nature, but not the same. So according to my initial observation, the altitude could easily only be a few miles.

After several hours of searching on Google, I came up with a logical explanation. The night this happened, it was Christmas Eve around the world. I discovered through my online search that around the world, there is a custom of launching fire lanterns on festive occasions. These fire lanterns are, in their most simple form, candles supported under paper bags. They lift off the ground from the heated air of the candles, much like a miniature hot air balloon. In a mild breeze, such as we had in Tequesta, Florida that evening, the candles, suspended under the paper bags, would sway from left to right as the wind blows. Indeed, upon further research, that was the explanation for what I observed. But it only became evident to me by researching the facts with an open mind. Until such time, those objects were UFOs. Not because they came from another world, but because I could not identify them right away.

I immediately updated my Youtube posting to reflect this new information. But for a few hours, I was caught up in the UFO craze since I had seen and recorded flying objects that I could, at least for a few hours, not identify. Now I never confirmed that these objects were indeed floating bag lanterns, but learning such things exist, are made all over the world and launched on festive holidays definitely seemed to be the most plausible explanation. Looking at the video now, as small as these objects appear on screen, it is clearly evident they display every characteristic of a fire lantern.

The lesson I learned from my own experience carries over to probably many other observations and recordings made of any UFO. The most plausible explanation, even for those UFOs not yet proven as to what they are, is most likely some sort of Earth bound flying object. Sometimes these objects are proven later to have been a secret military aircraft, a formation of ultralight aircraft (reported as UFOs some years ago over New York State) or some other wild flying contraption designed to purposely stimulate observers into thinking they witnessed a UFO. There are and always will be those individuals who like to create a hoax. With enough research, just as in my own case, UFOs have always been proven to have been some sort of earthly based phenomenon. No one has ever revealed an honest to goodness alien or spacecraft from another world. The closest we've come to that is meteorites.

What about the most famous UFO sighting of all times, the supposed crash of an alien spacecraft in Roswell, New Mexico in the summer of 1947? Why was the Air Force out to the crash site so fast to retrieve the debris and why did they try to dismiss it as the crash of a weather balloon? The most plausible explanation is this. The government did not want to disclose any possible information of the secret aircraft they were working on at the time. After all, we had just come out of World War Two after dropping two atomic bombs on Japan. We wanted to make sure if war ever broke out again, we would once again have every military advantage over our enemy. Everything was highly secretive to prevent other countries from discovering or stealing our secret research.

In order to throw off anyone from suspecting the crash of an exotic military plane, the government's best plausible story at the time was a crashed weather balloon. When the public didn't believe this, and instead came up with the fact that our government was covering up the crash of a UFO, they had the perfect alibi. They didn't even have to create it. We did that in our own minds. Because if they had come out and said the crash was an alien spacecraft, the public would have thought the government was pulling a fast one over on us.

Think about it for a moment. This came at a time when the US government was very busy designing advanced military aircraft in secret. Why so secret? Well, our government didn't want other governments to have any of the technological advances we created.

Why do they still hide information about the Roswell sighting today, so many years later? We the public will never know what secret technologies our government has, and with good reason. We don't want other nations to have the advantages we designed.

Perhaps we have developed some kind of nuclear blast-resistant metal alloy that can line an airplane underneath its outer skin, out of the sight of any prying eyes. Remember, we had just dropped an atomic bomb on Japan at that time, and we didn't want any other nation to be able to develop a nuclear bomb without us having some sort of protection from its radiation. It's just a theory, of course, but my thought is there is something of that magnitude our government still finds worthy of hiding. And what better way to throw people off track than by having them believe there's a UFO and an alien or two in hiding somewhere in a government lab?

When I look into the night sky, I see in my mind beyond what my eyes can show me. I see worlds out there waiting for the first group of its scientists and engineers to discover our radio signals. Most of our high band radio and TV signals drift past our atmosphere and travel forever into the voids of space, only to serve as a homing beacon for anyone with the ability to detect them. Perhaps our first proof of intelligent life beyond Earth will be in the form of radio and TV signals from another world. Let's hope our latest TV shows don't scare them.

The End